Contents

Some words are shown in bold, **like this**. You can find out what they mean by looking in the glossary.

What are magnets?

People use magnets every day. Magnets can pull towards each other. They can also push each other away. Magnets can also push or pull other objects.

⬆ Some toys use magnets.

Investigate

Magnets

Charlotte Guillain

www.raintreepublishers.co.uk
Visit our website to find out
more information about
Raintree books.

To order:
☎ Phone 0845 6044371
🖹 Fax +44 (0) 1865 312263
🖥 Email myorders@raintreepublishers.co.u

Customers from outside the UK please telephone +44 1865 312262

Raintree is an imprint of Capstone Global Library Limited,
a company incorporated in England and Wales having its
registered office at 7 Pilgrim Street, London, EC4V 6LB – Registered
company number: 6695582

Edited by Sarah Shannon, Catherine Clarke, and Laura Knowles
Designed by Joanna Hinton-Malivoire, Victoria Bevan,
    and Hart McLeod
Picture research by Liz Alexander and Rebecca Sodergren
Production by Duncan Gilbert
Originated by Chroma Graphics (Overseas) Pte. Ltd
Printed and bound in China by Leo Paper Group

ISBN 978 0 431 93278 1 (hardback)
12 11 10 09 08
10 9 8 7 6 5 4 3 2 1

ISBN 978 1 406 26603 0 (paperback)
13 12
10 9 8 7 6 5 4 3 2 1

British Library Cataloguing in Publication Data
Guillain, Charlotte
    Magnets. - (Investigate)

A full catalogue record for this book is available from the
British Library.

Acknowledgements
We would like to thank the following for permission to reproduce
photographs: ©Alamy pp. **4** (Friedrich Saurer), **13** (ELC); ©Corbis
pp. **11**, **29** (Sean Justice), **23** (Mark Ralston/Reuters), **25**, **26** Paul
Seheult: Eye Ubiquitous); ©Getty Images pp. **24** (Andersen Ross/
Blend Images), **28** (Joseph Van Os/Riser); ©istockphoto pp. **14**
(Matthew Cole), **18** (Ina Peters); ©Pearson Education Ltd. pp.
5, **17** (Tudor Photography), **6-10**, **12**, **15**, **16**, **19-21**, **30** (Lord and
Leverett 2007); ©Science Photo Library p. **22** (Jeremy Walker);
©Ufuk ZIVANA/istockphoto p. **26**.
Cover photograph of magnet letters on a blackboard
reproduced with permission of © Alamy (blickwinkel).

Every effort has been made to contact copyright holders of
material reproduced in this book. Any omissions will be rectified in
subsequent printings if notice is given to the publishers.

Disclaimer

Magnets come in many different shapes and sizes. They can be used in many ways.

How magnets work

Magnets use a **force** that can pull other objects towards them. This force is called **magnetism**. A force is a push or a pull. When a magnet pulls objects we say it **attracts** them. Magnets do not attract all objects.

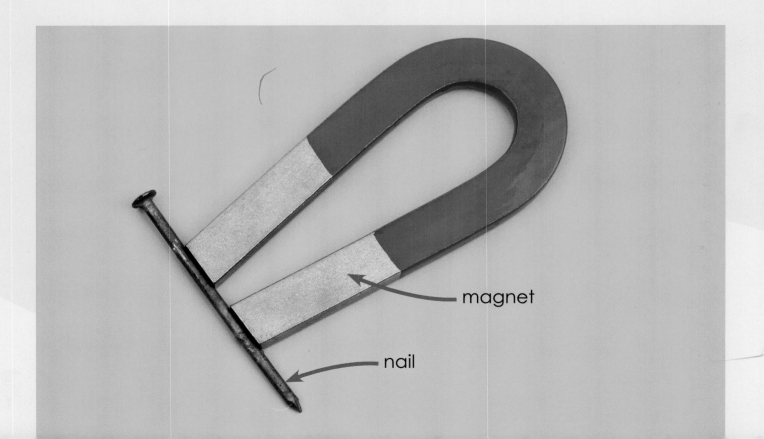

magnet

nail

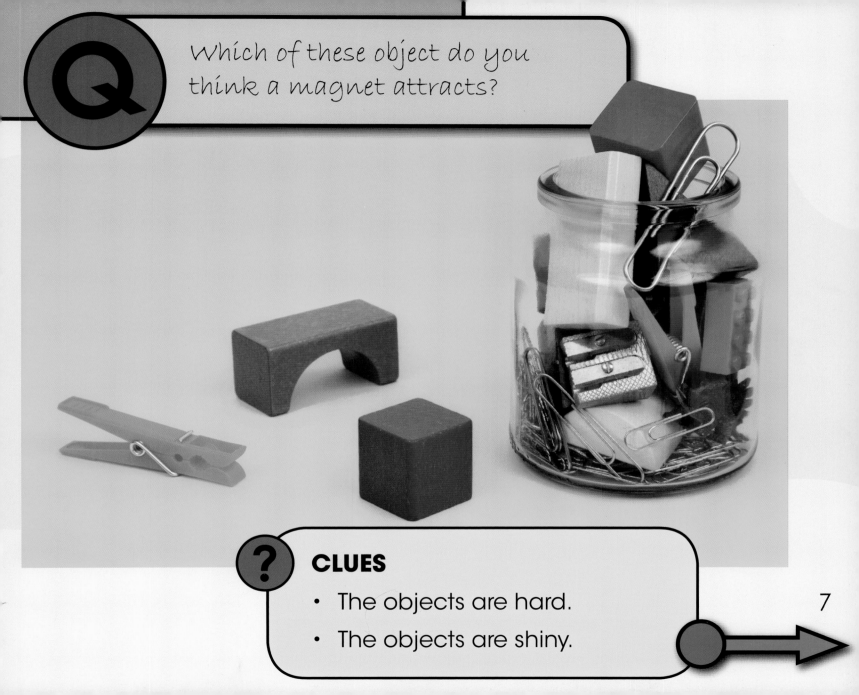

Q Which of these object do you think a magnet attracts?

? **CLUES**
- The objects are hard.
- The objects are shiny.

7

A magnet attracts metal objects such as paperclips. Magnets do not attract plastic, wood, glass, or any other materials.

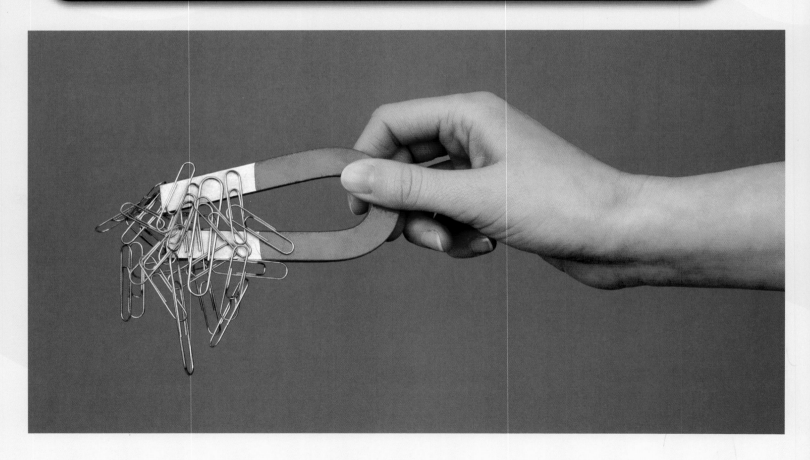

Magnets do not attract all metals. Magnets only attract objects made of certain metals such as iron, steel, and nickel. Cars, paper clips, fridges, washing machines, aeroplanes, and many other things are made of steel or iron.

A magnet would not attract this can because it is not made of iron.

9

Magnets have two ends or sides. These are called **poles**. A magnet has a north pole and a south pole. The north pole of one magnet attracts the south pole of another magnet.

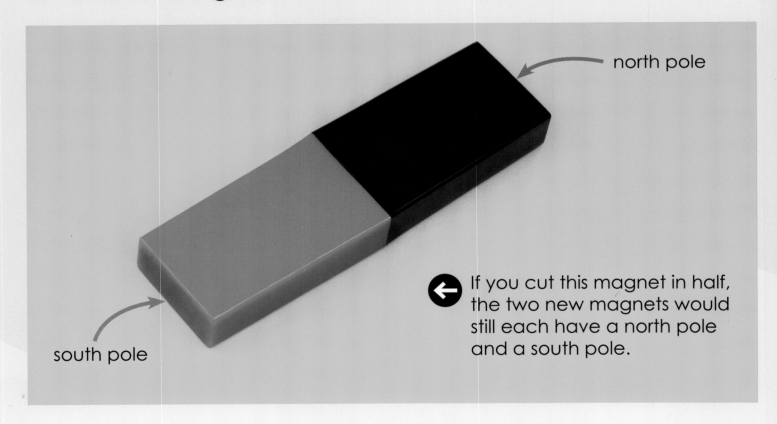

north pole

south pole

If you cut this magnet in half, the two new magnets would still each have a north pole and a south pole.

Q What happens if you put the south pole of one magnet against the south pole of another magnet?

 CLUE

• What is the opposite of pull?

A The south pole of one magnet pushes away the south pole of another magnet. It **repels** the other magnet.

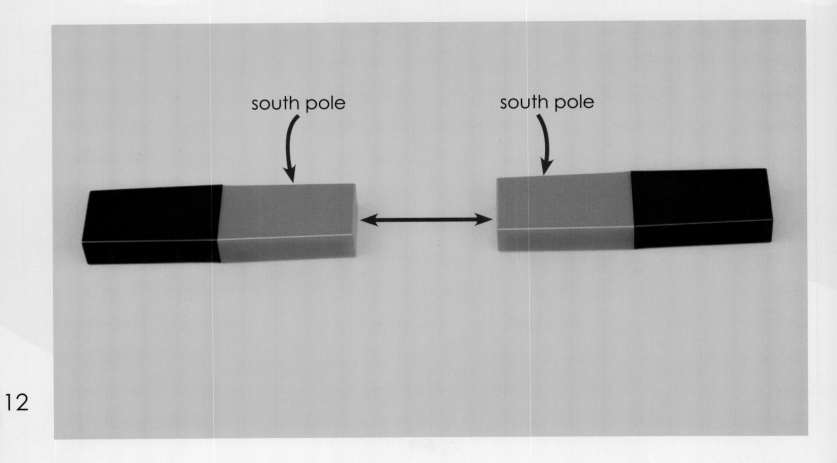

south pole   south pole

The north pole of one magnet will also repel the north pole of another magnet. Opposite poles attract. Poles that are alike repel each other.

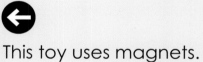
This toy uses magnets.

13

How strong are magnets?

A magnet can only pull objects that are inside its **magnetic field**. The magnetic field is the distance that a magnet's pull can reach. A magnet pulls most strongly at its **poles**.

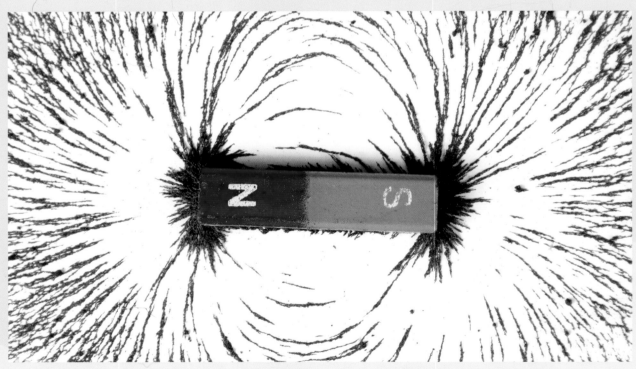

⬆ The iron filings around this magnet show its magnetic field. There are more filings at the ends of the magnets because this is where the pull is strongest.

The further the magnet is away from the paper clips, the less able it is to pull them.

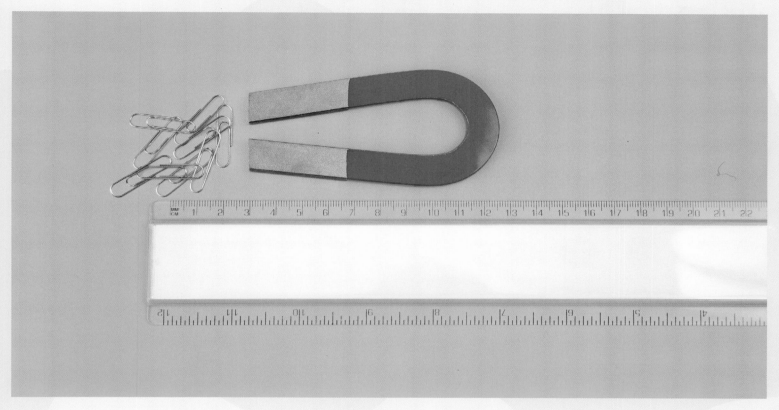

The pull is strongest when objects are close to the magnet. When objects are further away, it is harder for the magnet to pull them.

Some magnets have a stronger **magnetic field** than others. A magnet cannot pull objects that are too far away.

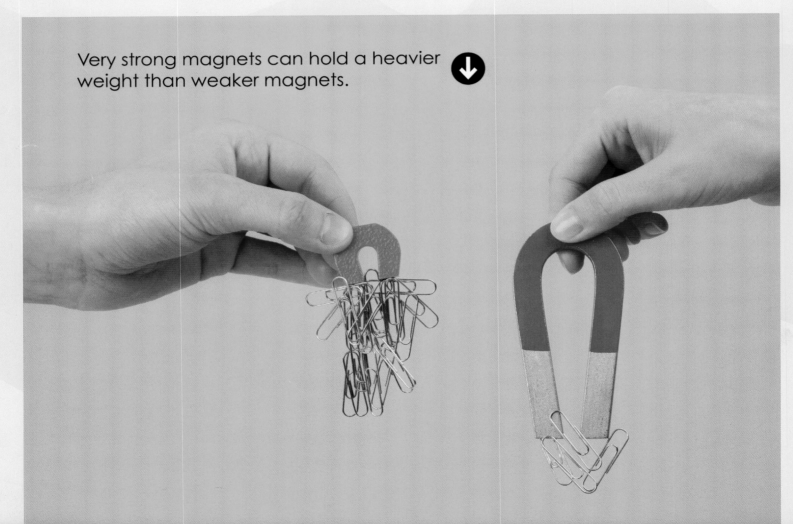

Very strong magnets can hold a heavier weight than weaker magnets.

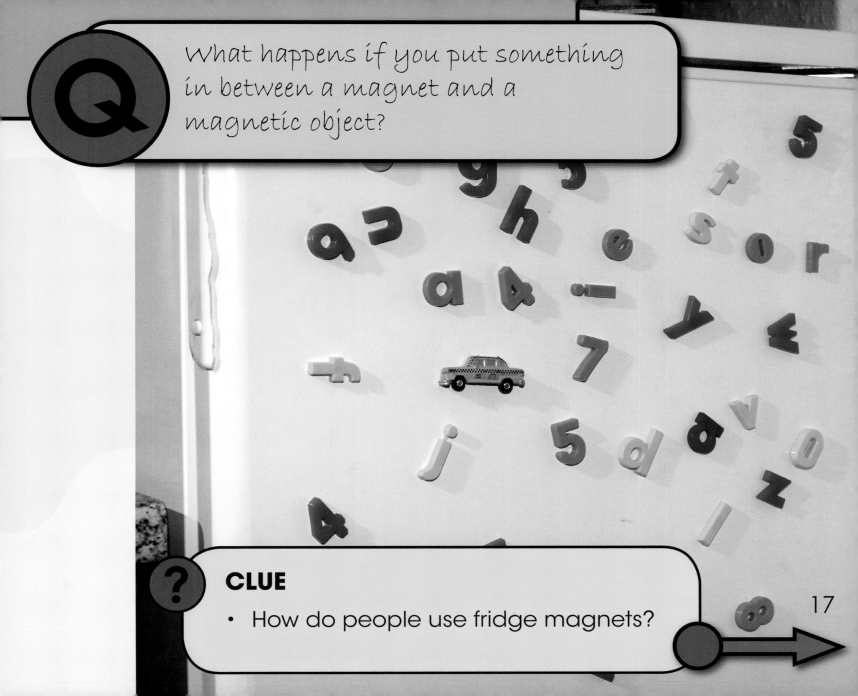

Q What happens if you put something in between a magnet and a magnetic object?

CLUE

• How do people use fridge magnets?

17

A A magnet can **attract** iron objects through other materials, such as paper. The magnetic pull can travel through other materials. But if the magnet is too weak, the magnetic pull will not reach through the other material.

This frog magnet can hold this sheet of paper against a fridge.

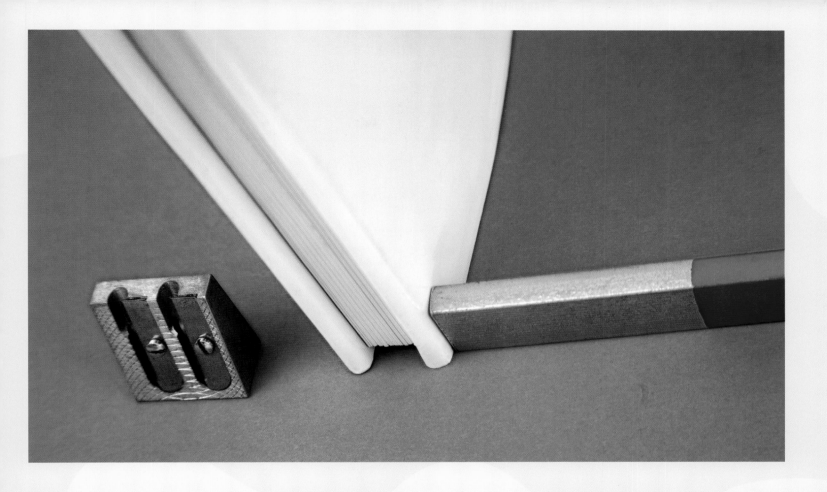

If the material in between is too thick, the magnetic pull will not reach an iron object. If you put too many pieces of paper under a fridge magnet, it will fall off the fridge.

Ways we use magnets

We use magnets in many ways at home. The doors on fridges, microwave ovens, and freezers have magnets around them to keep the door shut tightly.

Many toys use magnets. There are magnets in some tin openers and screwdrivers. There are even magnets in electric motors, CD players, computers, and telephones.

This fishing game uses magnets to pick up fish.

Magnets can be used to sort out rubbish. A large magnet **attracts** the iron and steel so it can be used again or **recycled**.

High-speed trains use magnets. There are magnets on the train and in the track that **repel** each other. This makes the train float above the track, so the ride is less bumpy. This means it can travel faster.

People use a magnetic **compass** to find the right direction. People use a magnetic compass when they are hiking.

Q This is a magnetic compass. Where do you think the magnet is?

? **CLUE**
- Can you see the north and south poles of the magnet?

A The needle on the compass is a magnet. The needle shows us which way is north. This helps us to find the right direction.

needle

26

Magnetic compasses work because the centre (core) of Earth is magnetic. There is hot melted iron in Earth's core.

core

Some animals have a **magnetic sense**. This sense helps them to find the right direction.

Migrating birds use magnetic sense to find their way.

Magnets are amazing! We use them all the time in our everyday lives. Scientists are finding new ways to use magnets all the time. Do you use anything that is magnetic?

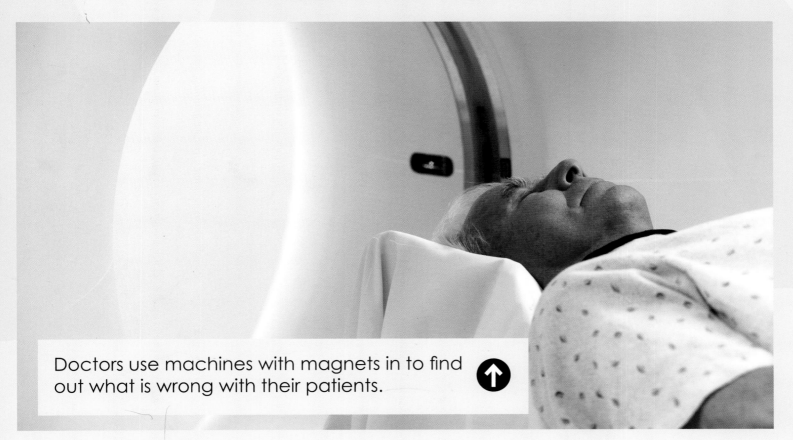

Doctors use machines with magnets in to find out what is wrong with their patients. ↑

Checklist

➟ A magnet can pull towards (**attract**) other magnets.

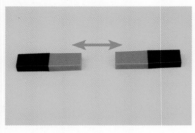

➟ A magnet can also push away (**repel**) other magnets.

➟ A magnet can attract some metals, including iron, steel, and nickel.

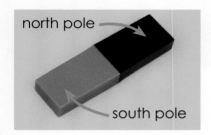

north pole

south pole

➟ A magnet has a north **pole** and a south pole.

30

Glossary

attract pull towards

compass tool used to find the right direction

force push or pull

magnetic field distance that a magnet's pull can reach

magnetic sense special sense some animals have to find their way

magnetism force that magnets use

poles opposite ends or sides of a magnet

recycle use again

repel push away

Index